23.05.19

D0530359

the moon

365 reflections

WITHDRAWN FROM STORE
DUBLIN CITY PUBLIC LIBRARIES

An Hachette UK Company
www.hachette.co.uk

First published in Great Britain in 2018 by Pyramid,
a division of Octopus Publishing Group Ltd
Carmelite House
50 Victoria Embankment
London EC4Y 0DZ
www.octopusbooks.co.uk

Copyright © Octopus Publishing Group Ltd 2018

All rights reserved. No part of this work may be reproduced or utilized
in any form or by any means, electronic or mechanical, including
photocopying, recording or by any information storage and retrieval
system, without the prior written permission of the publisher.

ISBN 978-0-75373-311-0

A CIP catalogue record for this book is available from the British Library

Printed and bound in China

10 9 8 7 6 5 4 3 2 1

Publisher: Lucy Pessell
Designer: Lisa Layton
Editor: Sarah Vaughan
Contributing Editor: Anna Bowles
Production Controller: Grace O'Byrne

WITHDRAWN FROM STOCK
DUBLIN CITY PUBLIC LIBRARIES

the
moon
365 reflections

In the time that man has been here
there is one thing that all have seen
and gazed upon in wonder. Luna.

Christopher Paul Flateau, photographer

The moon hung low in the sky like a yellow skull. From time to time a huge misshapen cloud stretched a long arm across and hid it.

Oscar Wilde, *The Picture of Dorian Gray*

Ancient cultures around the world worshipped the Moon as a divinity. Most Moon deities tended to be female, reflecting its traditional association with fertility.

We choose to go to the Moon in this decade and do the other things, not because they are easy, but because they are hard.

John F. Kennedy, Address at Rice University, September 12 1962

That's one small step for a man, one giant leap for mankind.

Neil Armstrong, the first man on the Moon

The Owl and the Pussy-cat

The Owl and the Pussy-cat went to sea
  In a beautiful pea-green boat,
They took some honey, and plenty of money,
  Wrapped up in a five-pound note.
The Owl looked up to the stars above,
  And sang to a small guitar,
"O lovely Pussy! O Pussy, my love,
  What a beautiful Pussy you are,
      You are,
      You are!
What a beautiful Pussy you are!"

Pussy said to the Owl, "You elegant fowl!
  How charmingly sweet you sing!
O let us be married! too long we have tarried:
  But what shall we do for a ring?"
They sailed away, for a year and a day,
  To the land where the Bong-Tree grows
And there in a wood a Piggy-wig stood
  With a ring at the end of his nose,
      His nose,
      His nose,
  With a ring at the end of his nose.

"Dear Pig, are you willing to sell for one shilling
  Your ring?" Said the Piggy, "I will."
So they took it away, and were married next day
  By the Turkey who lives on the hill.
They dined on mince, and slices of quince,
  Which they ate with a runcible spoon;
And hand in hand, on the edge of the sand,
  They danced by the light of the moon,
      The moon,
      The moon,
They danced by the light of the moon.

Edward Lear

Blessed is the Moon; it goes but it comes back again.

Samoan proverb

She used to tell me that
a full moon was when
mysterious things happen
and wishes come true.

Shannon A. Thompson, *November Snow*

So imagine that the lovely moon is playing just for you, everything makes music if you really want it to.

Giles Andreae, *Giraffes Can't Dance*

From "Night"

The moon, like a flower
In heaven's high bower,
With silent delight
Sits and smiles on the night.

William Blake

Where, indeed, does the moon not look well? What is the scene, confined or expansive, which her orb does not hallow?

Charlotte Brontë, *Villette*

In the Norse pantheon, the God of the Moon is Máni. Every night he is chased across the sky by a wolf called Hati.

*E*veryone is a moon,
and has a dark side
which he never shows
to anybody.

Mark Twain

The Moon will guide
you through the night
with her brightness,
but she will always
dwell in the darkness,
in order to be seen.

Shannon L. Alder

I am alone now, truly alone, and absolutely isolated from any known life. I am it. If a count were taken, the score would be three billion plus two over on the other side of the moon, and one plus God knows what on this side.

Michael Collins,
Carrying the Fire: An Astronaut's Journey

The moon in all her
immaculate purity
hung in the sky, laughing
at this world of dust.
She congratulated me for
my carefully considered
manoeuvres and invited me to
share in her eternal solitude.

Shan Sa, *Empress: A Novel*

The Man in the Moon
is a figure traditionally
seen on the surface of
the Moon. The image
is actually made up
by large, dark areas of
lava. Some cultures see
a rabbit or a woman
carrying a baby.

From "The Moon of Ramadân"

The sunset melts upon the Nile,
The stony desert glows,
Beneath heaven's universal smile,
One burning damask rose;
And like a Peri's pearly boat,
No longer than a span,
Look, faint on fiery sky afloat,
The Moon of Ramadân.

Mathilde Blind

It's a strange coincidence that the Sun and the Moon look the same size from Earth. The Sun is 400 times bigger than the Moon – but also 400 times further away.

In western astrology the Moon represents the emotional side of the self. The Moon's location in a person's birth chart reveals their creative side, where ideas and great passion arise.

The moon is a loyal companion.
It never leaves. It's always there,
watching, steadfast, knowing us in
our light and dark moments, changing
forever just as we do. Every day it's a
different version of itself. Sometimes
weak and wan, sometimes strong and
full of light. The moon understands
what it means to be human. Uncertain.
Alone. Cratered by imperfections.

Tahereh Mafi, *Shatter Me*

From "Ode to the Moon"

Pale Goddess of the witching hour;

Blest Contemplation's placid friend;

Oft in my solitary bow'r,

I mark thy lucid beam

From thy crystal car descend,

Whitening the spangled heath, and limpid sapphire stream.

Mary Darby Robinson

An English legend states that if Christmas falls on the day of a dark Moon – when the Moon is not visible – the following year's harvest will be bountiful.

The surface
of the Moon
has about the
same area as
the continent
of Africa.

I don't know if
there are men on
the Moon, but if
there are they must
be using the earth as
their lunatic asylum.

George Bernard Shaw

The Moon doesn't go through phases. Our perspective of the Moon goes through phases. No matter what the calendar says, the Moon is always full. Regardless of someone's opinion, perspective, or inability to see it as whole and complete, the Moon is unapologetically full. I find wisdom and strength in this truth.

Dr. Steve Maraboli

Whenever I gaze up at the Moon I feel like I'm on a time machine. I am back to that precious pinpoint of time, standing on the foreboding – yet beautiful – Sea of Tranquility. I could see our shining blue planet Earth poised in the darkness of Space.

Buzz Aldrin

The Moon moves
slowly, but it gets
across the town.

Ghanaian proverb

There is something haunting in the light of the moon; it has all the dispassionateness of a disembodied soul, and something of its inconceivable mystery.

Joseph Conrad, *Lord Jim*

Some ancient British peoples believed that a waxing Moon on Christmas night meant a good harvest the following autumn, but a waning Moon would bring a poor one.

In Chinese folklore,
the goddess of the
Moon, Chang'e, was
banished to the rock in
the sky after stealing
immortality elixir
from her husband, the
archer Hou Yi.

Her antiquity in preceding and surviving succeeding tellurian generations: her nocturnal predominance: her satellitic dependence: her luminary reflection: her constancy under all her phases, rising and setting by her appointed times, waxing and waning: the forced invariability of her aspect: her indeterminate response to inaffirmative interrogation: her potency over effluent and refluent waters: her power to enamour, to mortify, to invest with beauty, to render insane, to incite to and aid delinquency: the tranquil inscrutability of her visage: the terribility of her isolated dominant resplendent propinquity: her omens of tempest and of calm: the stimulation of her light, her motion and her presence: the admonition of her craters, her arid seas, her silence: her splendour, when visible: her attraction, when invisible.

James Joyce, *Ulysses*

Light from the
Moon takes about
a second and a
half to reach Earth.

A lunar month is
measured from new
Moon to new Moon.
It lasts about 29.5 days.

Phases of the Moon

Once upon a time I heard
That the flying moon was a Phoenix bird;
Thus she sails through windy skies,
Thus in the willow's arms she lies;
Turn to the East or turn to the West
In many trees she makes her nest.
When she's but a pearly thread
Look among birch leaves overhead;
When she dies in yellow smoke
Look in a thunder-smitten oak;
But in May when the moon is full,
Bright as water and white as wool,
Look for her where she loves to be,
Asleep in a high magnolia tree.

Elinor Morton Wylie

I feel a little like
the moon who took
possession of you for
a moment and then
returned your soul to you.
You should not love me.
One ought not to love the
moon. If you come too
near me, I will hurt you.

Anaïs Nin, *Delta of Venus*

From "The Sleeper"

At midnight, in the month of June,
I stand beneath the mystic moon.

Edgar Allan Poe

The Moon is moving
approximately 3.8 cm
(1.5 inches) further
away from the Earth
every year.

I suppose we shall soon travel by air-vessels; make air instead of sea voyages; and at length find our way to the Moon, in spite of the want of atmosphere.

Lord Byron

I never really thought about how when I look at the moon, it's the same moon as Shakespeare and Marie Antoinette and George Washington and Cleopatra looked at.

Susan Beth Pfeffer, *Life As We Knew It*

What was supposed to be so special about a full moon? It was only a big circle of light. And the dark of the moon was only darkness. But halfway between the two, when the moon was between the worlds of light and dark, when even the moon lived on the edge... maybe then a witch could believe in the moon.

Terry Pratchett, *Witches Abroad*

The ancient Inca people believed that a jaguar attacked and ate the Moon during eclipses. This was supposed to explain why it often turned blood red during a total eclipse. To drive the predator away, they would shake spears at the Moon, and make a lot of noise.

When a finger points to the Moon, the imbecile looks at the finger.

Chinese proverb

A blue Moon is a second full Moon that appears within a calendar month.

With freedom, books, flowers and the moon, who could not be happy?

Oscar Wilde, *De Profundis*

Eclipses of the Moon occur when Earth's shadow blocks the Sun's light, which otherwise reflects off the Moon.

On an Eclipse of the Moon

Struggling, and faint, and fainter didst thou wane,

O Moon! and round thee all thy starry train

Came forth to help thee, with half-open eyes,

And trembled every one with still surprise,

That the black Spectre should have dared assail

Their beauteous queen and seize her sacred veil.

Walter Savage Landor

The sky was a midnight-blue, like warm, deep, blue water, and the Moon seemed to lie on it like a water-lily, floating forward with an invisible current.

Willa Cather, *One of Ours*

In ancient Chinese belief, the Sun is symbolic of the fraternal (male, yang) aspect of guidance, and the Moon stands as the maternal (female, yin) influence.

There are hundreds of named craters on the Moon. Most of them are named after scientists and explorers.

The Hunter's Moon

The Hunter's Moon rides high,
High o'er the close-cropped plain;
Across the desert sky
The herded clouds amain
Scamper tumultuously,
Chased by the hounding wind
That yelps behind.

The clamorous hunt is done,
Warm-housed the kennelled pack;
One huntsman rides alone
With dangling bridle slack;
He wakes a hollow tone,
Far echoing to his horn
In clefts forlorn.

The Hunter's Moon rides low,
Her course is nearly sped.
Where is the panting roe?
Where hath the wild deer fled?
Hunter and hunted now
Lie in oblivion deep:
Dead or asleep.

Mathilde Blind

To buy a cake... to howl at the Moon... to know true happiness... I am happy.

C. JoyBell C., philosopher

The Moon! Artemis! the great goddess of the splendid past of men! Are you going to tell me she is a dead lump?

D. H. Lawrence, *The Posthumous Papers of D. H. Lawrence*

It was lunar symbolism that enabled man to relate and connect such heterogeneous things as: birth, becoming, death, and ressurection; the waters, plants, woman, fecundity, and immortality; the cosmic darkness, prenatal existence, and life after death, followed by the rebirth of the lunar type ("light coming out of darkness"); weaving, the symbol of the "thread of life," fate, temporality, and death; and yet others[...]We may even speak of a metaphysics of the moon, in the sense of a consistent system of "truths" relating to the mode of being peculiar to living creatures, to everything in the cosmos that shares in life, that is, in becoming, growth and waning, death and ressurrection.

Mircea Eliade, *The Sacred and the Profane:*
The Nature of Religion

From "Honour's Martyr"

The moon is full this winter night;
The stars are clear though few;
And every window glistens bright
With leaves of frozen dew.

The sweet moon through your lattice gleams,
And lights your room like day;
And there you pass, in happy dreams,
The peaceful hours away!

Emily Brontë

The moon had been observing
the earth close-up longer than
anyone. It must have witnessed all
of the phenomena occurring – and
all of the acts carried out – on this
earth. But the moon remained silent;
it told no stories. All it did was
embrace the heavy past with a cool,
measured detachment. On the moon
there was neither air nor wind. Its
vacuum was perfect for preserving
memories unscathed. No one could
unlock the heart of the moon.

Haruki Murakami, *1Q84*

Moonrise

And who has seen the moon, who has not seen
Her rise from out the chamber of the deep,
Flushed and grand and naked, as from the chamber
Of finished bridegroom, seen her rise and throw
Confession of delight upon the wave,
Littering the waves with her own superscription
Of bliss, till all her lambent beauty shakes towards us
Spread out and known at last, and we are sure
That beauty is a thing beyond the grave,
That perfect, bright experience never falls
To nothingness, and time will dim the moon
Sooner than our full consummation here
In this odd life will tarnish or pass away.

D. H. Lawrence

The Moon takes about 27 days to orbit the Earth.

Early scientists thought the dark patches on the Moon might be oceans of water. In fact they are made up of pools of hardened lava. Billions of years ago, the Moon had active volcanoes.

A Hymn to the Moon

Thou silver deity of secret night,
Direct my footsteps through the woodland shade;
Thou conscious witness of unknown delight,
The Lover's guardian, and the Muse's aid!
By thy pale beams I solitary rove,
To thee my tender grief confide;
Serenely sweet you gild the silent grove,
My friend, my goddess, and my guide.
E'en thee, fair queen, from thy amazing height,
The charms of young Endymion drew;
Veil'd with the mantle of concealing night;
With all thy greatness and thy coldness too.

Lady Mary Wortley Montagu

Moonlight is sculpture;
sunlight is painting.

Nathaniel Hawthorne

The Greek deity Artemis was goddess
of the Moon, and goddess of the hunt.
The Romans knew her as Diana.

The Moon rotates around on its own axis in exactly the same time it takes to orbit the Earth, meaning the same side is always facing the Earth. The side facing away from Earth has only been seen from spacecraft.

Ten years ago the Moon was an inspiration to poets and an opportunity for lovers. Ten years from now it will be just another airport.

Emmanuel G. Mesthene

Sonnet XLIV

Press'd by the Moon, mute arbitress of tides,
While the loud equinox its power combines,
The sea no more its swelling surge confines,
But o'er the shrinking land sublimely rides.
The wild blast, rising from the Western cave,
Drives the huge billows from their heaving bed;
Tears from their grassy tombs the village dead,
And breaks the silent sabbath of the grave!
With shells and sea-weed mingled, on the shore
Lo! their bones whiten in the frequent wave;
But vain to them the winds and waters rave;
They hear the warring elements no more:
While I am doom'd—by life's long storm opprest,
To gaze with envy on their gloomy rest.

Charlotte Smith

The Moon is friend
for the lonesome to
talk to.

Carl Sandburg

New Moon

The new moon, of no importance
lingers behind as the yellow sun glares and is gone beyond the sea's edge;
earth smokes blue;
the new moon, in cool height above the blushes,
brings a fresh fragrance of heaven to our senses.

D. H. Lawrence

Don't let's ask for the Moon!
We have the stars!

Olive Higgins Prouty, *Now, Voyager*

If the Moon is with you, you need not care about the stars.

African proverb

In ancient Egypt, the Moon was called Hathor-Tefnut. The first name was the Full Moon, and the second one the New. Hathor was represented by a girl, Tefnut by a lion.

Light died in the
west. Night and tears
took the Nation. The
star of Water drifted
among the clouds
like a murderer softly
leaving the scene of
the crime.

Terry Pratchett, *Nation*

The ancient Maya believed that the Moon was a young woman and the Sun a brave hunter. They fell in love and ran away together. The girl's grandfather became angry and had her killed, but dragonflies collected her body and blood and put them in thirteen hollow stumps.

Meanwhile, the Sun had been looking for his lover for thirteen days. On the thirteenth day he found the stumps. Twelve of them had given life to insects and snakes, which came out and filled the world. But from the thirteenth the Moon emerged – she had come back to life.

From "The Moon"

Beautiful Moon, with thy silvery light,
Thou seemest most charming to my sight;
As I gaze upon thee in the sky so high,
A tear of joy does moisten mine eye.

Beautiful Moon, with thy silvery light,
Thou cheerest the Esquimau in the night;
For thou lettest him see to harpoon the fish,
And with them he makes a dainty dish.

Beautiful Moon, with thy silvery light,
Thou cheerest the fox in the night,
And lettest him see to steal the grey goose away
Out of the farm-yard from a stack of hay.

William Topaz McGonagall

Some people believe that when you first see a crescent Moon for the month you should take all your spare coins out of your pocket, and put them in the other pocket. This is said to bring you good luck.

In Aztec mythology, Coyolxauhqui, meaning "golden bells", was the Moon goddess. She encouraged her four hundred brothers and sisters to kill their mother, the Earth goddess. As punishment, her head was cut off and thrown into the sky to form the Moon.

In presence of the Moon
nobody sees stars.

Amit Kalantr

A supermoon – a particularly
large Moon, as seen from Earth
– happens when the full Moon
coincides with the Moon's closest
approach to Earth in its orbit.

From "The Moon and the Yew Tree"

The moon is my mother.
She is not sweet like Mary.

Sylvia Plath

The moon does not fight. It attacks no one. It does not worry. It does not try to crush others. It keeps to its course, but by its very nature, it gently influences. What other body could pull an entire ocean from shore to shore? The moon is faithful to its nature and its power is never diminished.

Ming-Dao Deng, *Everyday Tao: Living with Balance and Harmony*

There is nothing
you can see that is
not a flower; there
is nothing you can
think that is not
the Moon.

Matsuo Basho

Is the Moon Tired?

Is the moon tired? she looks so pale
Within her misty veil:
She scales the sky from east to west,
And takes no rest.

Before the coming of the night
The moon shows papery white;
Before the dawning of the day
She fades away.

Christina Rossetti

"We must strive to be like the Moon." An old man in Kabati repeated this sentence often... the adage served to remind people to always be on their best behavior and to be good to others. [S]he said that people complain when there is too much sun and it gets unbearably hot, and also when it rains too much or when it is cold. But, no one grumbles when the moon shines. Everyone becomes happy and appreciates the Moon in their own special way.

Ishmael Beah, *A Long Way Gone: Memoirs of a Boy Soldier*

The ancident Mesopotamians believed that lunar eclipses were demons attacking the Moon. Eclipses were considered unlucky for the king, so when one was expected, a substitute king was installed. Once the danger was over, the real king returned to the throne.

The Australian saying
"to go between the Moon
and the milkman" means
to do an overnight flit,
often to avoid creditors.

Moonrise

I awoke in the Midsummer not to call night, in the white and the walk of the morning:

The moon, dwindled and thinned to the fringe of a finger-nail held to the candle,

Or paring of paradisaïcal fruit, lovely in waning but lustreless,

Stepped from the stool, drew back from the barrow, of dark Maenefa the mountain;

A cusp still clasped him, a fluke yet fanged him, entangled him, not quit utterly.

This was the prized, the desirable sight, unsought, presented so easily,

Parted me leaf and leaf, divided me, eyelid and eyelid of slumber.

Gerard Manley Hopkins

And all the insects
ceased in honour of
the moon.

Jack Kerouac, *Lonesome Traveler*

Only 59% of the Moon's surface is visible from Earth.

The surface
temperature on the
Moon varies from
-233 to 123 °C
(-384 to 253 °F).

The Moon Maiden's Song

Sleep! Cast thy canopy
Over this sleeper's brain,
Dim grow his memory,
When he wake again.

Love stays a summer night,
Till lights of morning come;
Then takes her winged flight
Back to her starry home.

Sleep! Yet thy days are mine;
Love's seal is over thee:
Far though my ways from thine,
Dim though thy memory.

Love stays a summer night,
Till lights of morning come;
Then takes her winged flight
Back to her starry home.

Ernest Christopher Dowson

Neil Armstrong and Buzz Aldrin were famously the first men to walk on the Moon in 1969. But there was a third astronaut on the mission, Michael Collins, who stayed in orbit 121 km (75 miles) above the surface.

The Moon

Thy beauty haunts me heart and soul,
Oh, thou fair Moon, so close and bright;
Thy beauty makes me like the child
That cries aloud to own thy light:
The little child that lifts each arm
To press thee to her bosom warm.

Though there are birds that sing this night
With thy white beams across their throats,
Let my deep silence speak for me
More than for them their sweetest notes:
Who worships thee till music fails,
Is greater than thy nightingales.

William Henry Davies

Many solemn nights Blond
moon, we stand and marvel...
Sleeping our noons away.

Basho Matsuo

When you fly to the
Moon you don't need
a rocket. You just
need the imagination!

Anthony T. Hincks

I ignored your aura but it grabbed me by the hand, like the Moon pulled the tide, and the tide pulled the sand.

Talib Kweli

No matter how bleak and black her existence became, the familiar sight of the moon restored something within her, small as it was – like tiny fluttering wings of flame beating back the darkness.

Shona Moyce, *Immisceo Taken*

from "The Light O' the Moon"

What the Snow Man Said

The Moon's a snowball. See the drifts
Of white that cross the sphere.
The Moon's a snowball, melted down
A dozen times a year.

Yet rolled again in hot July
When all my days are done
And cool to greet the weary eye
After the scorching sun.

The moon's a piece of winter fair
Renewed the year around,
Behold it, deathless and unstained,
Above the grimy ground!

It rolls on high so brave and white
Where the clear air-rivers flow,
Proclaiming Christmas all the time
And the glory of the snow!

Vachel Lindsay

Astronaut Neil Armstrong used to tell deliberately unfunny jokes about the Moon. When people looked baffled, he followed up with… "Ah, I guess you had to be there."

The Moon isn't round, but egg-shaped. It looks spherical from Earth because one of the smaller ends, with a circular silhouette, faces us.

Sleeping Out: Full Moon

They sleep within. . . .
I cower to the earth, I waking, I only.
High and cold thou dreamest, O queen, high-dreaming and lonely.

We have slept too long, who can hardly win
The white one flame, and the night-long crying;
The viewless passers; the world's low sighing
With desire, with yearning,
To the fire unburning,
To the heatless fire, to the flameless ecstasy! . . .

Helpless I lie.
And around me the feet of thy watchers tread.
There is a rumour and a radiance of wings above my head,
An intolerable radiance of wings. . . .

All the earth grows fire,
White lips of desire
Brushing cool on the forehead, croon slumbrous things.
Earth fades; and the air is thrilled with ways,
Dewy paths full of comfort. And radiant bands,

The gracious presence of friendly hands,
Help the blind one, the glad one, who stumbles and strays,
Stretching wavering hands, up, up, through the praise

Of a myriad silver trumpets, through cries,
To all glory, to all gladness, to the infinite height,
To the gracious, the unmoving, the mother eyes,
And the laughter, and the lips, of light.

Rupert Brooke

Sometimes the night can
be your best therapist.
For the Moon is free, and
always there to listen.

A. Y. Greyson

The Hupa tribe of Native Americans believed that the Moon had twenty wives and a lot of pet mountain lions and snakes. When he didn't bring them enough food, they attacked him and made him bleed. The eclipse would end when his wives came to protect him.

The Moon has much weaker gravity than Earth, due to its smaller mass, so you would weigh about one sixth of your weight on Earth.

The moon is
nothing But a
circumambulating
aphrodisiac Divinely
subsidized to provoke
the world Into a
rising birth-rate.

Christopher Fry,
The Lady's Not for Burning

Aim for the Moon. If you
miss, you may hit a star.

W. Clement Stone

The Moon and the Sea

Whilst the moon decks herself in Neptune's glass
And ponders over her image in the sea,
Her cloudy locks smoothing from off her face
That she may all as bright as beauty be;
It is my wont to sit upon the shore
And mark with what an even grace she glides
Her two concurrent paths of azure o'er,
One in the heavens, the other in the tides:
Now with a transient veil her face she hides
And ocean blackens with a human frown;
Now her fine screen of vapour she divides
And looks with all her light of beauty down;
Her splendid smile over-silvering the main
Spreads her the glass she looks into again.

George Darley

The first men on the Moon left behind some carefully chosen objects: an olive branch-shaped gold pin, messages from 73 world leaders, a patch from the Apollo 1 mission that, during a training exercize, combusted and killed three American astronauts, and medals in honour of two Soviet astronauts who had died in flight.

Who says you cannot hold the moon in your hand? Tonight when the stars come out and the moon rises in the velvet sky, look outside your window, then raise your hand and position your fingers around the disk of light. There you go... That was easy!

Vera Nazarian, *The Perpetual Calendar of Inspiration*

A full moon is poison to some; they shut it out at every crevice, and do not suffer a ray to cross them; it has a chemical or magical effect; it sickens them. But I am never more free and royal than when the subtile celerity of its magic combinations, whatever they are, is at work.

Harriet Prescott Spofford, *The Amber Gods*

The Luiseño tribe of southern California believed that an eclipse signaled that the Moon was ill. It was tribe members' job to sing chants or prayers to bring it back to health.

Earth is unusual in that it only has one moon. The giant planet Jupiter has 53, and more continue to be discovered!

It enclosed us in its laceries as we watched the moon spill across the Atlantic like wine from an overturned glass. With the light all around us, we felt secret in that moon-infused water like pearls forming in the soft tissues of oysters.

Pat Conroy, *Beach Music*

The Moon has water
ice at both its poles,
left there by comets
that hit the surface.

White in the Moon the Long Road Lies

White in the moon the long road lies,
The moon stands blank above;
White in the moon the long road lies
That leads me from my love.

Still hangs the hedge without a gust,
Still, still the shadows stay:
My feet upon the moonlit dust
Pursue the ceaseless way.

The world is round, so travellers tell,
And straight though reach the track,
Trudge on, trudge on, 'twill all be well,
The way will guide one back.

But ere the circle homeward hies
Far, far must it remove:
White in the moon the long road lies
That leads me from my love.

A. E. Housman

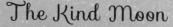

The Kind Moon

I think the moon is very kind
To take such trouble just for me.
He came along with me from home
To keep me company.

He went as fast as I could run;
I wonder how he crossed the sky?
I'm sure he hasn't legs and feet
Or any wings to fly.

Yet here he is above their roof;
Perhaps he thinks it isn't right
For me to go so far alone,
Tho' mother said I might.

Sara Teasdale

The Moon has no atmosphere, so it is unprotected from meteorites, which create the millions of craters that cover the surface.

Moonlight incites dark
passions like a cold flame,
making hearts burning with
the intensity of phosphorus.

Edogawa Rampo

New Moon

What have you got in your knapsack fair,
White moon, bright moon, pearling the air,
Spinning your bobbins and fabrics free,
Fleet moon, sweet moon, in to the sea?
Turquoise and beryl and rings of gold,
Clear moon, dear moon, ne'er to be sold?
Roses and lilies, romance and love,
Still moon, chill moon, swinging above?
Slender your feet as a white birds throat,
High moon, shy moon, drifting your boat
Into the murk of the world awhile,
Slim moon, dim moon, adding a smile.
Tender your eyes as a maiden's kiss,
Fine moon, wine moon, no one knows this,
Under the spell of your witchery,
Dream moon, cream moon, first he kissed me.

Zora Bernice May Cross

Monday is the day of silence,
day of the whole white mung bean,
which is sacred to the moon.

Chitra Banerjee Divakaruni, *The Mistress of Spices*

In the Norse pantheon, the God of the Moon is Mani. Every night he is chased across the sky by a wolf called Hati.

In South African Bushman legend, the Moon is a man who has angered the Sun. Every month the Moon reaches round prosperity, but the Sun's knife then cuts away pieces until finally only a tiny piece is left, which the Moon pleads should be left for his children. It is from this piece that the Moon gradually grows again to become full.

The Moon, our own, earthly Moon is bitterly lonely, because it is alone in the sky, always alone, and there is no one to turn to, no one to turn to it. All it can do is ache across the weightless airy ice, across thousands of versts, toward those who are equally lonely on earth, and listen to the endless howling of dogs.

Yevgeny Zamyatin, *A Story about the Most Important Thing*

The Moon is about
384,400 kilometres
(250,000 miles) from Earth.

The Young May Moon

The young May moon is beaming, love.
The glow-worm's lamp is gleaming, love.
How sweet to rove,
Through Morna's grove,
When the drowsy world is dreaming, love!
Then awake! -- the heavens look bright, my dear,
'Tis never too late for delight, my dear,
And the best of all ways
To lengthen our days
Is to steal a few hours from the night, my dear!

Now all the world is sleeping, love,
But the Sage, his star-watch keeping, love,
And I, whose star,
More glorious far,
Is the eye from that casement peeping, love.
Then awake! -- till rise of sun, my dear,
The Sage's glass we'll shun, my dear,
Or, in watching the flight
Of bodies of light,
He might happen to take thee for one, my dear.

Thomas Moore

The night walked
down the sky with the
Moon in her hand.

Frederic Lawrence Knowles

Look Down, Fair Moon

Look down, fair moon, and bathe this scene;
Pour softly down night's nimbus floods, on faces ghastly, swollen, purple;
On the dead, on their backs, with their arms toss'd wide,
Pour down your unstinted nimbus, sacred moon.

Walt Whitman

from "The Cat and the Moon"

The cat went here and there
And the moon spun round like a top,
And the nearest kin of the moon,
The creeping cat, looked up.
Black Minnaloushe stared at the moon,
For, wander and wail as he would,
The pure cold light in the sky
Troubled his animal blood.

W. B. Yeats

In Malawian legend, the morning star is Chechichani, a poor housekeeper who allows her husband the Moon to go hungry and starve. Puikani, the evening star, is a fine wife who feeds the Moon, thus bringing him back to life.

The Moon is essentially gray, no color. It looks like plaster of Paris, like dirty beach sand with lots of footprints in it.

James A. Lovell

I've never seen a Moon in the sky that, if it didn't take my breath away, at least misplaced it for a moment.

Colin Farrell

"Where did you live before you came here?" I asked. "The moon," he said smoothly. "We left because the place had no atmosphere."

Laurie Halse Anderson,
The Impossible Knife of Memory

The Moon looks upon many night flowers; the night flowers see but one Moon.

Jean Ingelow

Moonset

Idles the night wind through the dreaming firs,
That waking murmur low,
As some lost melody returning stirs
The love of long ago;
And through the far, cool distance, zephyr fanned.
The moon is sinking into shadow-land.

The troubled night-bird, calling plaintively,
Wanders on restless wing;
The cedars, chanting vespers to the sea,
Await its answering,
That comes in wash of waves along the strand,
The while the moon slips into shadow-land.

O! soft responsive voices of the night
I join your minstrelsy,
And call across the fading silver light
As something calls to me;
I may not all your meaning understand,
But I have touched your soul in shadow-land.

Emily Pauline Johnson

The Crazed Moon

Crazed through much child-bearing
The moon is staggering in the sky;
Moon-struck by the despairing
Glances of her wandering eye
We grope, and grope in vain,
For children born of her pain.

Children dazed or dead!
When she in all her virginal pride
First trod on the mountain's head
What stir ran through the countryside
Where every foot obeyed her glance!
What manhood led the dance!

Fly-catchers of the moon,
Our hands are blenched, our fingers seem
But slender needles of bone;
Blenched by that malicious dream
They are spread wide that each
May rend what comes in reach.

W. B. Yeats

Go out of the house to see the
Moon, and 'tis mere tinsel:
it will not please as when
its light shines upon your
necessary journey.

Ralph Waldo Emerson

It is a beautiful and
delightful sight to behold
the body of the Moon.

Galileo Galilei

Among the Xhosa of South Africa it used to be believed that beyond the sea was a vast pit filled with new moons ready for use. Each new Moon really was new!

In our village, folks say God crumbles up the old Moon into stars.

Aleksandr Solzhenitsyn,
One Day in the Life of Ivan Denisovich

From "To the Moon – Rydal"

Queen of the stars!--so gentle, so benign,
That ancient Fable did to thee assign,
When darkness creeping o'er thy silver brow
Warned thee these upper regions to forego,
Alternate empire in the shades below--
A Bard, who, lately near the wide-spread sea
Traversed by gleaming ships, looked up to thee
With grateful thoughts, doth now thy rising hail
From the close confines of a shadowy vale.

William Wordsworth

When I admire the wonders of a sunset or the beauty of the Moon, my soul expands in the worship of the creator.

Mahatma Gandhi

The Moon has a molten core, like the Earth.

Anyhow, I took every stitch of clothing off and got out of bed. And I got down on my knees on the floor in the white moonlight. The heat was off and the room must have been cold, but I didn't feel cold. There was some kind of special something in the moonlight and it was wrapping my body in a thin, skintight film. At least that's how I felt. I just stayed there naked for a while, spacing out, but then I took turns holding different parts of my body out to be bathed in the moonlight. I don't know, it just seemed like the most natural thing to do.

Haruki Murakami, *The Wind-Up Bird Chronicle*

There are nights when
the wolves are silent and
only the Moon howls.

George Carlin

Mister Moon, Mister Moon,
You're out too soon;
The sun is still in the sky.
Go back to bed and cover up your head,
And wait till the day goes by!

Traditional Nursery Rhyme

The moon climbed out of the ravine, blue, skinny, as if it had been fed on nothing but skimmed milk. It climbed out, and quickly slithered up and up along the finest thread — away from trouble, and on the very top it huddled, crouching on thin legs.

Yevgeny Zamyatin,
The Protectress of Sinners

The Moon is slipping from our grasp at a rate of about 1.5 inches a year. In another two billion years it will have receded so far that it won't keep us steady and we will have to come up with some other solution, but in the meantime you should think of it as much more than just a pleasant feature in the night sky.

Bill Bryson,
A Short History of Nearly Everything

For the Tartars of Central Asia, the Moon, known as the Queen of Life and Death, was dualistic, representing both the forces of creation and destruction.

From now on we live in a world where man has walked on the Moon. It's not a miracle; we just decided to go.

Tom Hanks

The Moon is the reflection of your heart and moonlight is the twinkle of your love.

Debasish Mridha

from "The Light O' the Moon"
The Beggar Speaks

"What Mister Moon Said to Me."

Come, eat the bread of idleness,
Come, sit beside the spring:
Some of the flowers will keep awake,
Some of the birds will sing.

Come, eat the bread no man has sought
For half a hundred years:
Men hurry so they have no griefs,
Nor even idle tears:

They hurry so they have no loves:
They cannot curse nor laugh —
Their hearts die in their youth with neither
Grave nor epitaph.

My bread would make them careless,
And never quite on time —
Their eyelids would be heavy,
Their fancies full of rhyme:

Each soul a mystic rose-tree,
Or a curious incense tree:
Come, eat the bread of idleness,
Said Mister Moon to me.

Vachel Lindsay

The moon is
just another
kind of clock.

Kelli Russell Agodon,
Hourglass Museum

Tonight or every night if
you wish you can have
a very distinguished guest
from the space: Just open
your curtain at night, then
the Moon will visit you!

Mehmet Murat Ildan

I like to think that the
Moon is there even if
I am not looking at it.

Albert Einstein

Hey Diddle Diddle,
The cat and the fiddle,
The cow jumped over the Moon.
The little dog laughed,
To see such fun,
And the dish ran away with the spoon.

Traditional English Nursery Rhyme

The least of us is improved by the things done by the best of us, because if we are not able to land at least we are able to follow.

Walter Cronkite, during the CBS TV Moon Landing Coverage, July 20, 1969

Full Moon

around the Moon, so shining and full,
stars hide their luminous faces,
while she spreads her silvery light
over the vast Earth.

Sappho

The Roman Moon goddess Luna was said to inspire madness in non-believers. Her name is where we get the word "lunatic" from.

And he beholds the Moon; like a rounded fragment of ice filled with motionless light.

Gustave Flaubert,
The Temptation of St. Anthony

The Moon is a Painter

He coveted her portrait.
He toiled as she grew gay.
She loved to see him labor
In that devoted way.

And in the end it pleased her,
But bowed him more with care.
Her rose-smile showed so plainly,
Her soul-smile was not there.

That night he groped without a lamp
To find a cloak, a book,
And on the vexing portrait
By moonrise chanced to look.

The color-scheme was out of key,
The maiden rose-smile faint,
But through the blessed darkness
She gleamed, his friendly saint.

The comrade, white, immortal,
His bride, and more than bride —
The citizen, the sage of mind,
For whom he lived and died.

Vachel Lindsay

Those are the same stars,
and that is the same
Moon, that look down upon
your brothers and sisters, and
which they see as they look
up to them, though they are
ever so far away from us, and
each other.

Sojourner Truth

The surface of the Moon
is not smooth, uniform and
precisely spherical as a
great number of philosophers
believe it to be, but is uneven,
rough, and full of cavities
and prominences, being
not unlike the face of the
Earth, relieved by chains of
mountains and deep valleys.

Galileo Galilei

We'll go
back to the
Moon by
not learning
anything new.

Burt Rutan

from "The Rime of the Ancient Mariner"

The moving Moon went up the sky,
And no where did abide:
Softly she was going up,
And a star or two beside –

Her beams bemocked the sultry main,
Like April hoar-frost spread;
But where the ship's huge shadow lay,
The charméd water burnt alway
A still and awful red.

Samuel Taylor Coleridge

She'd never really looked at the moon, never really seen how intricate the etchings on its yellowy silver surface. Bowl of a spoon in candlelight. When she'd looked a long time – I see the moon, and the moon sees me – a glimmering ring like a rainbow materialized at the rim. In the memory she still retained, as clear as a framed snapshot, a portrait worn in a locket, Saga stared at the moon that way for hours, and it kept her company, it kept her sane, it kept her in one piece, it kept her alive. It was proof, fact, patience, faith.

Julia Glass, *The Whole World Over*

The Waning Moon

And like a dying lady, lean and pale,
Who totters forth, wrapped in a gauzy veil,
Out of her chamber, led by the insane
And feeble wanderings of her fading brain,
The moon arose up in the murky east,
A white and shapeless mass.

Percy Bysshe Shelley

Somewhere I'd heard, or invented perhaps, that the only pleasures found during a waning moon are misfortunes in disguise. Superstition aside, I avoid pleasure during the waning or absent moon out of respect for the bounty this world offers me. I profit from great harvests in life and believe in the importance of seasons.

Roman Payne, *Rooftop Soliloquy*

The Sumerian god of the Moon was known as Sin. His high priest was chosen from the royal family, and regarded as the god's spouse.

I'm convinced that before the year 2000 is over, the first child will have been born on the Moon.

Wernher von Braun,
pioneering rocket engineer, 1972

'Tis said that some people are moonstruck, we find,
   But the Man in the Moon must be out of his mind.
But it can't be for love for he's quite on his own,
No ladies to meet him by moonlight alone.
It can't be ambition, for rivals he's none,
At least he is only eclipsed by the sun,
But when drinking, I say, he is seldom surpassed,
For he always looks best when he's seen through a glass.

"Man in the Moon", traditional folk song from Norfolk, UK

The moon is the friendliest
of the celestial bodies,
after all, glowing warm and
white and welcoming, like
a friend who wants only to
know that all of us are safe
in our narrow worlds, our
narrow yards, our narrow,
well-considered lives. The moon
worries. We may not know how
we know that, but we know
it all the same: that the moon
watches, and the moon worries,
and the moon will always love
us, no matter what.

Seanan McGuire,
Down Among the Sticks and Bones

The loveliest
faces are to be
seen by moonlight,
when one sees half
with the eye and
half with the fancy.

Iranian proverb

The phases of the
Moon are: new
Moon, crescent, first
quarter, waxing
gibbous, full Moon,
waning gibbous, last
quarter, crescent…
and back to new.

Maybe the Moon just lost herself gazing too long at the brilliance of the Sun and that's how she got her glow. And maybe we're made of the same mysterious sort of magic that makes us magnify and mirror whatever we look at the most.

Cristen Rogers

A kiss on the beach
when there is a full
moon is the closest
thing to heaven.

H. Jackson Brown Jr.,
Life's Little Instruction Book

You know the Portrait in the Moon

You know that Portrait in the Moon—
So tell me who 'tis like—
The very Brow—the stooping eyes—
A fog for—Say—Whose Sake?

The very Pattern of the Cheek—
It varies—in the Chin—
But—Ishmael—since we met—'tis long—
And fashions—intervene—

When Moon's at full—'Tis Thou—I say—
My lips just hold the name—
When crescent—Thou art worn—I note—
But—there—the Golden Same—

And when—Some Night—Bold—slashing Clouds
Cut Thee away from Me—
That's easier—than the other film
That glazes Holiday—

Emily Dickinson

In September 1950, a real blue Moon was seen over Scotland and the north of England. It was almost certainly caused by wind-borne particles thrown into the atmosphere by an enormous wildfire raging in Canada.

Don't tell me the sky is the limit, there are footprints on the Moon!

Dorothy Parker

What the Moon Saw

Two statesmen met by moonlight.
Their ease was partly feigned.
They glanced about the prairie.
Their faces were constrained.
In various ways aforetime
They had misled the state,
Yet did it so politely
Their henchmen thought them great.
They sat beneath a hedge and spake
No word, but had a smoke.
A satchel passed from hand to hand.
Next day, the deadlock broke.

Vachel Lindsay

Autumn is coming. For as long as I can remember, I've talked to the moon. Asked her for her guidance. There's something deeply spiritual about her waxing and waning. She wears a new dress every evening, yet she's always herself. And she's always there.

Stephanie Perkins, *Lola and the Boy Next Door*

From "The Use of the Moon"

The moon is a silver pin-head vast
That holds the heaven's tent-hangings fast.

William R. Alger

The next time you stand on a beach at night, watching the moon's bright path across the water, and the conscious of the moon-drawn tides, remember that the moon itself may have been born of a great tidal wave of earthly substance, torn off into space. And remember if the moon was formed in this fashion, the event may have had much to do with shaping the ocean basins and the continents as we know them.

Rachel Carson, marine biologist, *The Sea Around Us*

O swear not by the moon, th'inconstant moon
That monthly changes in her circle orb
Let that thy love prove likewise variable.

William Shakespeare, *Romeo and Juliet*

Mawu, the Moon, is the supreme god of the Fon people of Benin. She is seen as an old mother who lives in the west, and brings cooler temperatures to the hot, dusty land.

I think we're going to the Moon because it's in the nature of the human being to face challenges. It's by the nature of his deep inner soul... we're required to do these things just as salmon swim upstream.

Neil Armstrong

Sun adores the body
Moon romances your soul...

Shonali Dey, *The Essence of Eternal Happiness*

To the Sad Moon

With how sad steps, O Moon, thou climb'st the skies!
How silently, and with how wan a face!
What! May it be that even in heavenly place
That busy archer his sharp arrows tries?
Sure, if that long-with-love-acquainted eyes
Can judge of love, thou feel'st a lover's case:
I read it in thy looks; thy languished grace
To me, that feel the like, thy state descries.
Then, even of fellowship, O Moon, tell me,
Is constant love deemed there but want of wit?
Are beauties there as proud as here they be?
Do they above love to be loved, and yet
Those lovers scorn whom that love doth possess?
Do they call "virtue" there— ungratefulness?

Sir Philip Sidney

The Moon is the
first milestone on
the road to the stars.

Arthur C. Clarke

For me at age 11, I had a pair of binoculars and looked up to the Moon, and the Moon wasn't just bigger, it was better. There were mountains and valleys and craters and shadows. And it came alive.

Neil deGrasse Tyson

The blank face of the moon looked down wistfully on the pair and tried to lean in just a little closer.

S. E. Grove, *Glass Sentence*

In Inuit tradition, the Moon god Anningan chases his sister Malina, the Sun goddess, across the sky, but forgets to eat as he goes, so he gets much thinner.

from "Hudibras, Part II"

The moon pull'd off her veil of light
That hides her face by day from sight,
(Mysterious veil, of brightness made,
That's both her lustre and her shade,)
And in the lantern of the night
with shining horns hung out her light;
For darkness is the proper sphere,
Where all false glories use t' appear.

Samuel Butler

What was most significant about the lunar voyage was not that men set foot on the Moon but that they set eye on the earth.

Norman Cousins

The Moon teaches us that darkness can't hide the beauty of life if we know how to reflect beauty.

Debasish Mridha

Humans have dreamed of going to the Moon for millennia. Some of the first known stories were Lucian's *Icaromenippus* and *True Story*, written in the 2nd century AD.

Footprints left on the Moon by Apollo astronauts will remain visible for at least 10 million years because there is no erosion on the Moon.

The Chinese considered the moon to be yin, feminine and full of negative energy, as opposed to the sun that was yang and exemplified masculinity. I liked the moon, with its soft silver beams. It was at once elusive and filled with trickery, so that lost objects that had rolled into the crevices of a room were rarely found, and books read in its light seemed to contain all sorts of fanciful stories that were never there the next morning.

Yangsze Choo, *The Ghost Bride*

I don't go along with going to Moon first to build a launch pad to go to Mars. We should go to Mars from Earth orbit. We have already been to the Moon; we've already practiced.

Wally Schirra

The Sun and Moon Must Make Their Haste

The Sun and Moon must make their haste—
The Stars express around
For in the Zones of Paradise
The Lord alone is burned—

His Eye, it is the East and West—
The North and South when He
Do concentrate His Countenance
Like Glow Worms, flee away—

Oh Poor and Far—
Oh Hindred Eye
That hunted for the Day—
The Lord a Candle entertains
Entirely for Thee—

Emily Dickinson

The moon was
a fang in the
lightning sky.

Alexander Maksik,
You Deserve Nothing

I've seen it when the crescent moon shone bright on a cold, dark night. The darker the night, the brighter smile.

Anusha Atukorala,
Glimpses of Light

The Mamaiurans, an Amazonian tribe, believe that at the beginning of time there were so many birds in the sky that their wings blocked out the sunlight. Two heroes, Iae and his brother Kuat, wearied of the darkness and decided to force Urubutsin, king of the birds, to share the daylight with humans. The brothers hid, and waited for Urubutsin to land, when Kuat grabbed his leg. Unable to escape, the king was forced to make an agreement that the birds would share the light with the Mamiurans. Since then, day has alternated with night, and Kuat is seen as representing the Sun and Iae the Moon.

You develop an instant global consciousness, a people orientation, an intense dissatisfaction with the state of the world, and a compulsion to do something about it. From out there on the Moon, international politics look so petty. You want to grab a politician by the scruff of the neck and drag him a quarter of a million miles out and say, "Look at that, you son of a bitch."

Edgar D. Mitchell

Fragment: "To the Moon"

Art thou pale for weariness
Of climbing Heaven, and gazing on the earth,
Wandering companionless
Among the stars that have a different birth,--
And ever changing, like a joyless eye
That finds no object worth its constancy?

Percy Bysshe Shelley

Moon and Sea

You are the moon, dear love, and I the sea:
The tide of hope swells high within my breast,
And hides the rough dark rocks of life's unrest
When your fond eyes smile near in perigee.
But when that loving face is turned from me,
Low falls the tide, and the grim rocks appear,
And earth's dim coast-line seems a thing to fear.
You are the moon, dear one, and I the sea.

Ella Wheeler Wilcox

The full Moon is symbolic of the height of power, the peak of clarity, fullness and obtainment of desire.

Our Moon is the fifth largest moon in the solar system.

I x Chel, "Lady Rainbow", was the old Moon goddess in Maya mythology. She was the protector of weavers and women in childbirth, but depicted as a fierce old woman wearing a skirt with crossed bones and holding a serpent in her hand.

Even when the moon shrinks and disappears, it shows itself again gradually. When ancient people saw that eternal cycle of death and recovery, they prayed to the moon for their own rebirth. Rebirth. Will I be reborn? ... If I were reborn, what would I become?

Fuminori Nakamura, *The Kingdom*

The Man in the Moon came down too soon,
and asked his way to Norwich,
They sent him south and he burnt his mouth
By eating cold pease-porridge.

Traditional English rhyme

If you strive for the
Moon, maybe you'll
get over the fence.

James Wood

We walked on the
moon. We made
footprints somewhere
no one else had ever
made footprints, and
unless someone comes
and rubs them out,
those footprints will be
there forever because
there's no wind.

Frank Cottrel Boyce, *Cosmic*

Unlike the Earth,
the Moon has
no magnetic field.
Scientists are still
unsure of why this is.

How sweet the moonlight sleeps upon this bank!

William Shakespeare,
The Merchant of Venice

At a Lunar Eclipse

Thy shadow, Earth, from Pole to Central Sea,
    Now steals along upon the Moon's meek shine
In even monochrome and curving line
Of imperturbable serenity.

How shall I link such sun-cast symmetry
With the torn troubled form I know as thine,
That profile, placid as a brow divine,
With continents of moil and misery?

And can immense Mortality but throw
So small a shade, and Heaven's high human scheme
Be hemmed within the coasts yon arc implies?

Is such the stellar gauge of earthly show,
Nation at war with nation, brains that teem,
Heroes, and women fairer than the skies?

Thomas Hardy

It is no secret that the Moon
has no light of her own,
but is, as it were, a mirror,
receiving brightness from the
influence of the Sun.

Vitrivius

When a dog barks at the Moon,
then it is religion; but when he
barks at strangers, it is patriotism!

David Starr Jordan

Who is but pleased to watch the Moon on high

Who but is pleased to watch the Moon on high
Travelling where she from time to time enshrouds
Her head, and nothing loth her Majesty
Renounces, till among the scattered clouds
One with its kindling edge declares that soon
Will reappear before the uplifted eye
A Form as bright, as beautiful a moon,
To glide in open prospect through clear sky.
Pity that such a promise e'er should prove
False in the issue, that yon seeming space
Of sky should be in truth the stedfast face
Of a cloud flat and dense, through which must move
(By transit not unlike man's frequent doom)
The Wanderer lost in more determined gloom.

William Wordsworth

The Moon was but a Chin of Gold

The Moon was but a chin of gold
A night or two ago,
And now she turns her perfect face
Upon the world below.
Her forehead is of amplest blond;
Her cheek like beryl stone;
Her eye unto the summer dew
The likest I have known.
Her lips of amber never part;
But what must be the smile
Upon her friend she could bestow
Were such her silver will!
And what a privilege to be
But the remotest star!
For certainly her way might pass
Beside your twinkling door.
Her bonnet is the firmament,
The universe her shoe,
The stars the trinkets at her belt,
Her dimities of blue.

Emily Dickinson

The moon's a
crazy sweetheart.

Helen Humphreys, *Leaving Earth*

The Yongu Aboriginal people of Australia call the Moon Ngalindi. They believe that he was originally a fat, lazy man (the full Moon) who was punished by his wives, who chopped bits off him (the waning Moon). He managed to escape by climbing a tall tree to follow the Sun, but was mortally wounded and died (the new Moon). After remaining dead for three days he rose again (the waxing Moon) until after two weeks his wives attacked him again.

The flirty old moon eased his way across the warped and sooty floorboards and kissed my bare toes, turning my feet as luminous as the skin of cinema stars.

Cat Winters, *Emmeline*

Shadows on the Moon are far darker than those on Earth due to the lack of atmosphere (which normally diffuses light rays). Anything the Sun doesn't shine on directly is pitch black. Once an astronaut's foot steps into shadow, they can't see it any more.

There are deep craters at the poles of the Moon where the sunlight never reaches. These places have seen no light for over a billion years.

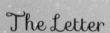

The Letter

Little cramped words scrawling all over the paper
Like draggled fly's legs,
What can you tell of the flaring moon
Through the oak leaves?
Or of my uncertain window and the bare floor
Spattered with moonlight?
Your silly quirks and twists have nothing in them
Of blossoming hawthorns,
And this paper is dull, crisp, smooth, virgin of loveliness
Beneath my hand.

I am tired, Beloved, of chafing my heart against
The want of you;
Of squeezing it into little inkdrops,
And posting it.
And I scald alone, here, under the fire
Of the great moon.

Amy Lovell

I'm a moon junkie.
Every time I look
at the Moon I feel less
alone and less afraid.
I tell my boys that
moonlight is a magic
blanket and the stars
above are campfires set
by friendly aliens.

Amy Poehler, *Yes Please*

Nights would have
been expressionless
had it not been for the
moon. The moon, I say,
is a mood.

Geetika Kohli, *Yonder*

Because the Moon has no atmosphere, no sound can be heard there and the sky always looks black. What will it profit this country if we put our man on the Moon by 1970 and at the same time you can't walk down Woodward Avenue in this city without fear of some violence?

Jerome Cavanagh

From "At Sunset"

Into the sunset's turquoise marge
The moon dips, like a pearly barge;
Enchantment sails through magic seas,
To fairyland Hesperides,
Over the hills and away.

Madison Cawein

The full Moon's light poured into the room like a stroke from a wide paintbrush.

Peter Hammarberg, *Antilla*

Immovable is my heart, like a mountain.
Irresistible is your beauty, like a moon.

Swami Cidananda Tirtha

Traditional Maori belief states that a woman called Rona was carrying a bucket of stream water home to her children when she tripped over a root in the darkness of the night. She cursed the Moon for not lighting her way. The Moon heard her remarks, and grabbed her and her bucket. Many people today see a woman with a bucket in the Moon, and the Maori say that when Rona upsets her bucket, it rains.

I believe that the only way that the human race is gonna survive is to start colonizing space and setting up colonies on the Moon, and then space stations.

Ace Frehley, KISS

The Moon is considered a relatively easy object to land humans on, everything else is much harder by orders of magnitude. It is the reason why we have not been to Mars and will likely never go there successfully with humans.

Steven Magee, astronomer

I kept staring at the moon. I'm not sure if its light was good or evil. I thought it might not be either. The moon just shines with the light of chaos. Mysteriously. Brightly. That must not be either good or evil. Just as the rules of this world are not all good.

Fuminori Nakamura, *The Kingdom*

Ah, Moon—and Star!

Ah, Moon—and Star!
You are very far—
But were no one
Farther than you—
Do you think I'd stop
For a Firmament—
Or a Cubit—or so?

I could borrow a Bonnet
Of the Lark—
And a Chamois' Silver Boot—
And a stirrup of an Antelope—
And be with you—Tonight!

But, Moon, and Star,
Though you're very far—
There is one—farther than you—
He—is more than a firmament—from Me—
So I can never go!

Emily Dickinson

Why doesn't anyone go to the moon anymore? What happened to our optimism?

Janet Turpin Myers, *Nightswimming*

Buzz Aldrin claims that he, not Neil Armstrong, spoke the first words on the Moon. Six hours before Armstrong stepped onto the surface, Aldrin said "Contact light" as the Eagle landing module touched down.

The Moon's weird though, right? It's there, and there, and then suddenly it's not. And it seems to be pretty far up. Is it watching us? If not, what is it watching instead? Is there something more interesting than us? Hey, watch us Moon! We may not always be the best show in the universe, but we try.

Welcome to Night Vale podcast, episode five, "The Shape in Grove Park"

From "To the Moon" – on the coast of Cumberland

Wanderer! that stoop'st so low, and com'st so near
    To human life's unsettled atmosphere;
Who lov'st with Night and Silence to partake,
So might it seem, the cares of them that wake;
And, through the cottage-lattice softly peeping,
Dost shield from harm the humblest of the sleeping;
What pleasure once encompassed those sweet names
Which yet in thy behalf the Poet claims,
An idolizing dreamer as of yore!--
I slight them all; and, on this sea-beat shore
Sole-sitting, only can to thoughts attend
That bid me hail thee as the SAILOR'S FRIEND;
So call thee for heaven's grace through thee made known
By confidence supplied and mercy shown,
When not a twinkling star or beacon's light
Abates the perils of a stormy night;
And for less obvious benefits, that find
Their way, with thy pure help, to heart and mind;
Both for the adventurer starting in life's prime;
And veteran ranging round from clime to clime,
Long-baffled hope's slow fever in his veins,
And wounds and weakness oft his labour's sole remains.

William Wordsworth

Diana

How like a Queen comes forth the lonely Moon
From the slow-opening curtains of the clouds,
Walking in beauty to her midnight throne!
The stars are veil'd in light; the ocean-floods,
And the ten thousand streams — the boundless woods,
The trackless wilderness — the mountain's brow,
Where Winter on eternal pinions broods—
All height, depth, wildness, grandeur, gloom, below,
Touched by thy smile, lone Moon! in one wide splendour glow.

George Croly

Six American flags have been left on the Moon, but they no longer show the Star and Stripes – they have been bleached to a blank white by the Sun's radiation.

We are going to
the Moon that
is not very far.
Man has so
much farther to
go within himself.

Anaïs Nin

Meditate. Live purely. Be quiet. Do your work with mastery. Like the Moon, come out from behind the clouds! Shine.

Buddha

In a myth of the Luyia people of Kenya in East Africa, the Moon and Sun were brothers. The Sun was jealous of the Moon, who was older, bigger and brighter, so he picked a fight. As the two wrestled, the Moon fell into the mud, and his brightness was dimmed. God finally made them stop fighting and kept them apart by ordering the sun to shine by day and the mud-spattered moon to shine by night to illuminate the world of witches and thieves.

When I look over my past, I see that the stages in my life are like the phases of the moon. I've had periods where I was the waxing gibbous: fat with wealth and success. There have been other seasons when my happiness was like the waning crescent and I watched my joy fade away slowly, merging with the atmosphere around me as if it never existed.

Amy Neftzger, *Conversations with the Moon*

He stared up at the moon, which looked like a giant hole in the sky, letting light through to the other side.

Sarah Addison Allen, *Garden Spells*

You may want to change the place or the direction of the Moon or you may do something more practical: You accept the Moon as it is!

Mehmet Murat Ildan

The elegance
of the Moon
teaches us
to be elegant
and kind.

Debasish Mridha

I promise to be an
excellent husband,
but give me a wife
who, like the Moon,
will not appear
every day in my sky.

Anton Chekhov

The Moon

The Moon has a face like the clock in the hall;
She shines on thieves on the garden wall,
On streets and fields and harbour quays,
And birdies asleep in the forks of the trees.

The squalling cat and the squeaking mouse,
The howling dog by the door of the house,
The bat that lies in bed at noon,
All love to be out by the light of the moon.

But all of the things that belong to the day
Cuddle to sleep to be out of her way;
And flowers and children close their eyes
Till up in the morning the sun shall arise.

Robert Louis Stevenson

The Moon is the only astronomical body other than Earth ever visited by human beings.

The last man to walk on the Moon was Gene Cernan on the Apollo 17 mission in 1972. Since then the Moon has only be visited by unmanned vehicles.

A Night Thought

Lo! where the Moon along the sky
Sails with her happy destiny;
Oft is she hid from mortal eye
Or dimly seen,
But when the clouds asunder fly
How bright her mien!
Far different we--a froward race,
Thousands though rich in Fortune's grace
With cherished sullenness of pace
Their way pursue,
Ingrates who wear a smileless face
The whole year through.
If kindred humours e'er would make
My spirit droop for drooping's sake,
From Fancy following in thy wake,
Bright ship of heaven!
A counter impulse let me take
And be forgiven.

William Wordsworth

Niagara

Seen on a night in
November
How frail
Above the bulk
Of crashing water hangs,
Autumn, evanescent, wan,
The moon.

Adelaide Crapsey

Mons Huygens is the tallest mountain on the Moon. At 4,700m (15,400 ft), it is just over half the height of Mount Everest.

The Moon is covered with dust that is as fine as flour, but very rough. It clings to everything, erodes astronauts' boots and acts like sandpaper on their visors. Astronauts get "Moon hay fever" from inhaling it.

A southern moon is a sodden moon, and sultry. When it swamps the fields and the rustling sandy roads and the sticky honeysuckle hedges in its sweet stagnation, your fight to hold on to reality is like a protestation against a first waft of ether.

Zelda Fitzgerald, *Save Me the Waltz*

"Why is it always such a surprise?" thinks Toby. "The moon. Even though we know it's coming. Every time we see it, it makes us pause, and hush."

Margaret Atwood, *MaddAddam*

The ancient Chinese believed that there were twelve moons, as there were twelve months in one year.

From "The Moon Stopped Behind the Lake"

The Moon stopped behind the lake
Seeming like an open window
Into a quiet, brightly-lit house
Where something unpleasant had happened.

Anna Akhmatova

She didn't quite know what the relationship was between lunatics and the moon, but it must be a strong one, if they used a word like that to describe the insane.

Paulo Coelho, *Veronika Decides to Die*

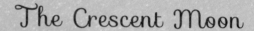

The Crescent Moon

Slipping softly through the sky
Little horned, happy moon,
Can you hear me up so high?
Will you come down soon?

On my nursery window-sill
Will you stay your steady flight?
And then float away with me
Through the summer night?

Brushing over tops of trees,
Playing hide and seek with stars,
Peeping up through shiny clouds
At Jupiter or Mars.

I shall fill my lap with roses
Gathered in the milky way,
All to carry home to mother.
Oh! what will she say!

Little rocking, sailing moon,
Do you hear me shout -- Ahoy!
Just a little nearer, moon,
To please a little boy.

Amy Lowell

One's shadow grows larger than life when admired by the light of the Moon.

Chinese proverb

I fancied my luck to be witnessing yet another full moon. True, I'd seen hundreds of full moons in my life, but they were not limitless. When one starts thinking of the full moon as a common sight that will come again to one's eyes ad-infinitum, the value of life is diminished and life goes by uncherished.

Roman Payne, *The Wanderess*

In Hindi myth, the Moon is said to be the storehouse of the elixir of immortality that only the gods can drink.

...on the darkest night, the maidens take their spindles down to the sea, to wash their wool. And the wool slips from the spindles into the water, and unravels in long ripples of light from the shore to the horizon, and there is the moon again, rising above the sea.

Mary Stewart, *The Moon-Spinners*

About fifty per cent of the soil on the Moon is glass, formed by rocks melting when they were struck by meteorites landing at high speed.

Tell me the story...
About how the sun loved the moon so much...
That she died every night...
Just to let him breathe...

Hanako Ishii

The first known photograph of the Moon was taken by a Dr. J. W. Draper in New York in 1840.

Moon and Earth is
the longest known
stable love affair
known to mankind.

Danny Maximus

who knows if the moon's
a balloon, coming out of a keen city
in the sky – filled with pretty people?

E. E. cummings

If I ever get to go to the Moon, I'll probably just stand on the Moon and go "Hmmm, yeah... fair enough... gotta go home now."

Noel Gallagher

As well as craters, NASA has detected about two hundred holes on the Moon's surface. It's thought that these may lead down to lava tubes – tunnels formed by lava flows millions of years ago. Future astronauts might be able to establish an underground base in these caverns.

Treading the soil of the Moon, palpating its pebbles, tasting the panic and splendor of the event, feeling in the pit of one's stomach the separation from terra... these form the most romantic sensation an explorer has ever known...

Vladimir Nabokov, *New York Times*, commenting on the arrival of the first man on the Moon

O, fluttering moon,
if only we
could hang a handle on you,
what a fan you would be!

Yamazaki Sōkan

The first recorded prediction of a human colony on the Moon was A Discourse Concerning a New World and Another Planet, by Bishop John Wilkins, published in 1638.

Legends of werewolves have been around for many hundreds of years, but the idea that the creature's transformation from man to beast is caused by the full Moon is actually a relatively recent notion.

During the 1950s, the USA considered detonating a nuclear bomb on the Moon. The secret project, under the harmless-sounding code name of "Project A119" was intended as a show of strength.

It is pleasant to sit in the green wood and watch the Sun in his chariot of gold, and the Moon in her chariot of pearl.

Oscar Wilde, *The Nightingale and the Rose*

from "The Light O' the Moon"
What the Old Horse said

The moon's a peck of corn. It lies
Heaped up for me to eat.
I wish that I might climb the path
And taste that supper sweet.

Men feed me straw and scanty grain
And beat me till I'm sore.
Some day I'll break the halter-rope
And smash the stable-door,

Run down the street and mount the hill
Just as the corn appears.
I've seen it rise at certain times
For years and years and years.

Vachel Lindsay

There was an Old Man of the Hague,
    Whose ideas were excessively vague;
He built a balloon to examine the moon,
That deluded Old Man of the Hague.

Edward Lear

The Outer Space Treaty, ratified by the United Nations in 1967, states that nobody owns the Moon.

Moon is the light from a lantern in heaven.

Munia Khan

I bet a fun thing would be to go way back in time to where there was going to be an eclipse and tell the cave men, "If I have come to destroy you, may the sun be blotted out from the sky." Just then the eclipse would start, and they'd probably try to kill you or something, but then you could explain about the rotation of the Moon and all, and everyone would get a good laugh.

Jack Handey

O Moon that rid'st the night to wake
Before the dawn is pale,
The hamadryad in the brake,
The Satyr in the vale,
Caught in thy net of shadows
What dreams hast thou to show?

Dr Gerald Gardner, *The Meaning of Witchcraft*

$\mathcal{T}$he ancient
    Chinese believed
that the Moon was
made of water.

The horns of the
crescent Moon always
point away from the
Sun, as the visible
crescent is the part
that is lit by the
Sun's rays.

...lifting my cup,
I asked the Moon
to drink with me...

Li Po

Are we not
acceptable, moon?
Are we not lovely
sitting together here,
I in my satin; he in
black and white?

Virginia Woolf, *The Waves*

from "The Light O' the Moon"
What the Hyena Said

The moon is but a golden skull,
  She mounts the heavens now,
And Moon-Worms, mighty Moon-Worms
Are wreathed around her brow.

The Moon-Worms are a doughty race:
They eat her gray and golden face.
Her eye-sockets dead, and molding head:
These caverns are their dwelling-place.

The Moon-Worms, serpents of the skies,
From the great hollows of her eyes
Behold all souls, and they are wise:
With tiny, keen and icy eyes,
Behold how each man sins and dies.

When Earth in gold-corruption lies
Long dead, the moon-worm butterflies
On cyclone wings will reach this place —
Yea, rear their brood on earth's dead face.

Vachel Lindsey

The man who has
seen the rising Moon
break out of the clouds
at midnight has been
present like an archangel
at the creation of light
and of the world.

Ralph Waldo Emerson

Three things
cannot be long
hidden: the Sun,
the Moon, and
the truth.

Buddha